Global Higher Education Technology Transfer Office Outreach & Marketing Activities

ISBN 1-57440-193-9

Library of Congress Control Number: 2011945254

TABLE OF CONTENTS

LIST OF TABLES

QUESTIONNAIRE

INTERNAL AND EXTERNAL MARKETING AND TECHNOLOGY DEVELOPMENT

1. How many invention disclosure reports did the office receive from faculty in the following years:
   a. 2009?
   b. 2010?
   c. And expected in 2011?
2. What percentage of the invention disclosure reports was the Office initially contacted by the Faculty member in the following years (as opposed to solicited by the Office itself):
   a. 2009?
   b. 2010?
   c. And planned in 2011?
3. What percentage resulted from strong encouragement by the technology transfer office in the following years:
   a. 2009?
   b. 2010?
   c. And planned in 2011?
4. Are any employees only responsible for internal marketing to faculty? If so, how many in the following years:
   a. 2009?
   b. 2010?
   c. And planned in 2011?
5. Are any employees responsible for only external marketing to Industry? How many in the following years:
   a. 2009?
   b. 2010?
   c. 2011?
6. How many professionals are dedicated to spin out companies only in the following years:
   a. 2009?
   b. 2010?

c. 2011?

7. Does the Office use an outside public relations firm or does it use the university public relations department?
    a. Outside public relations firm
    b. University public relations department

DEVELOPMENTS IN PUBLIC RELATIONS

8. How often do you send out releases? To whom do you send them? What results have you achieved and what do you recommend?
9. Do you use a PR firm? University PR? Both? Why? What do you recommend for peers?
10. Do you send electronic releases, paper releases or both? (choose all that apply)
    a. Electronic
    b. Paper
11. Have you used e-marketing extensively? Google ads, blog marketing, Facebook marketing, opt in email, search engine optimization or ezine advertising? If you have used any of these means what have been the results?
12. How much did the office spend on the following in 2010?
    a. Mass mailings
    b. Web advertising (Google ads/Facebook/other online marketing
    c. Printing of promotional materials
    d. Trade shows/fairs/Invention fairs – Travel and lodging costs
    e. Trade shows/fairs/Invention fairs – Booth & attendance fees
    f. Space ads in publications
    g. Website development
    h. Video and media productions

PARTNERSHIPS

13. Does the Office partner with any Foreign Universities? If so, how many in the following years:
    a. 2009?
    b. 2010?
    c. 2011?
14. Which countries does the office partner with?

15. Does the office partner with any Industry or company groups? How Many? What industries generally? Describe your efforts.
16. Does the office support special research institutes or organizations of technology interest? If so, how? With funding? Advisory services? Explain.
17. Does the office invest in university startup spinout companies with funding or management assistance? Does the office help arrange financing for private sector vendors in its technology?
18. Continued from previous question: Is there a special investment fund with office management and employees? How many?
19. How easy is it for your office to collaborate with other university outreach departments such as the business school, grants and contracts, university outreach, international programs? Are there budget constraints to collaboration?
20. Has your institution merged your office with others or otherwise re-organized the institutional relationship between your office and the Office of the General Counsel, the medical or business schools, grants and contracts or other offices?

PARTICIPANTS

Beth Israel Deaconess Medical Center
Boston University OTD
Cardiff University
Case Western Reserve University
City University London
Clemson University Research Foundation
Florida State University
Isis Innovation, University of Oxford
Johns Hopkins University
Joslin Diabetes Center
Michigan Technological University
National Jewish Health
North Carolina Agricultural & Technical State University
Ohio University
Oregon Health & Science University
Oregon State University
REWACO
Rush University Medical Center
The University of Toledo
UH Off of Tech Trans & Econ Dev
University of Aberdeen
University of Guelph
University of Louisville
University of Manitoba
University of Medicine & Dentistry of New Jersey
University of Michigan
University of North Dakota
University of Notre Dame
University of Surrey
University of the West of Scotland
Virginia Tech Intellectual Properties
Wake Forest University Baptist
Wichita State University

ABOUT THE AUTHORS

The authors are extremely active and knowledgeable with the business and economics of Global Higher Education and Technology Transfer Offices and always welcome comments, inquires, questions and special benchmarking projects for your organization.

Joseph R. Flicek is President of Blake Technologies a New York Corporation established in 1994. Joseph R. Flicek has over 25 years of management experience in all phases of business development, commercialization and intellectual property management in technology and healthcare industries. Mr. Flicek currently works with and advises universities and business on all phases of technology transfer and licensing issues in the United States and globally in Europe, Asia, and the Middle East.

Mr. Flicek established the New Ventures Office of the Science & Technology Ventures Office at Columbia University and served as Director from 1996 to 2002. The New Ventures Office help established over 30 companies. The Columbia portfolio companies have raised almost $1 billion in venture and public market financings, and have generated more than more than 1,500 jobs. Mr. Flicek's primary responsibilities were to work with faculty members, industry and venture capital groups to start new business ventures based on Columbia University intellectual property from the more than 7,000 faculty members and 2,500 laboratories. Columbia is the number one university with earnings of over $150 million per year in licensing revenues. Mr. Flicek served as Chief Executive Officer of M2M Imaging, an MRI Medical Device company, and SIPCOMM, a Voice-Over-IP company. Both companies were Columbia University technology spinouts.

James Moses, President of Primary Research Group, Inc, has been a consultant to a broad range of corporations and governments including McDonalds, Copyright Clearance Center, and the Government of Spain. Mr. Moses is the author of hundreds of industry studies and benchmarking reports for Primary Research Group and many other report publishers.

He is the former contributing editor of the American Management *Association's Growth Strategies* and a contributor to *Agenda, Business International, Multinational Business, The Journal of Commerce* and other publications.

A former economist for the US International Trade Commission, Mr. Moses holds an MA in international political economy from Columbia University's School of International & Public Affairs.

Contact the authors at PrimaryResearchGroup@gmail.com.

SUMMARY OF MAIN FINDINGS

Chapter 1: Internal and External Marketing and Technology Development

NUMBER OF INVENTION DISCLOSURE REPORTS FILED

Invention disclosure reports received by university technology transfer offices are expected to rebound again in 2011 by 5.2% over 2010 after declining 3.75% in 2010. This is based in part on a somewhat more hopeful economy. Invention disclosure reports received by university technology transfer offices declined by 3.75% in 2010 from 2009. This was to be expected from the tough economic conditions experienced globally in 2009 and 2010.

US universities on average generate about 68% more invention disclosures than universities abroad. The mean number of invention disclosures received in the US and Abroad in 2009 was 80, in 2010 was 77 and 81 are forecasted for 2011. US universities forecasted a mean of 91 invention disclosures in 2011 while foreign university forecasted a mean of 54 invention disclosures.

The maximum number of invention disclosures received in 2010 by one university was 342 and the minimum number received was 2 invention disclosure reports.

NUMBER OF FOREIGN PATENTS FILED

The offices in the sample filed a mean of 30.66 foreign patents in the past two years with a range of 0 to 325.

Foreign patent filings were highest in offices with a major concentration of activity in medical technology; these offices filed a mean of nearly 40 non-US patents in the past two years. As might be expected median foreign patent filings increased with office size, peaking at a mean of 40.5 with offices with 14 or more employees fte.

Chapter 2: Developments in Public Relations

TECHNOLOGY MARKETING EFFORTS

A shade more than three quarters of the survey participants send electronic press releases, though only a third of smaller universities and colleges did so, while nearly 89% of major universities used this form of marketing. Less than a third of those surveyed send out paper versions of press releases; more than 44% of non-USA survey participants did so, as did 75% of hospitals and 50% of those institutions focused on energy technology.

For the most part, university technology transfer offices do not seem to have heavily utilized online marketing techniques. In this respect they significantly lag other university administrative departments such as admissions and enrollment, alumni services, athletics, and others – for whom the use of such venues has become standard practice. It seems odd, even unseemly, that major university offices entrusted with licensing technology are relatively unlikely to use the major technological advances in marketing practices in the modern era. It may also suggest that cooperation with other centers of university marketing may generate some knowledge transfer.

Traditional marketing such as direct mail and other venues was also sparsely used. However, mean spending on website development in 2010 was $2,979.20.

Chapter 3: Partnerships

PARTNERSHIPS WITH FOREIGN INSTITUTIONS

The institutions in the sample had a mean of 1.78 partnerships with foreign institutions in 2009. Almost all partnerships were formed by non-USA institutions which had a mean of 4.63 foreign partnerships in technology development or marketing; US institutions had a mean of only 0.27. Colleges with a strong medical technology sector were far more likely than others to have international partnerships – they had a mean of 2.6 foreign partnerships in 2009.

The mean number of foreign partnerships rose to 2.17 in 2010, an increase of nearly 22.5%. Non USA institutions once again led this charge, increasing their mean number of foreign partnerships to 5.88 from 4.63. In 2011 the institutions sampled planned another major increase in their cooperation with foreign partners, expecting an increase to a mean of 2.63 partnerships (from 2.17) an increase of more than 20%. Non USA institutions expected to increase their mean number of partnerships to 7.13 in 2011.

TECHNOLOGY TRANSFER OFFICE MERGERS WITH OTHER ACADEMIC DEPARTMENTS

We asked the organizations sampled if they had merged their office or otherwise re-organized their institutional relationship with the University's Office of General Counsel, Grants and Contracts Office, or medical or business school, or their office that provides overlapping or similar services as the technology transfer office. More than 24% of survey participants had engaged in some form of merger or office re-organization to enhance economies of scale, reduce duplication of services or in some other way enhance cooperation while improving productivity in the provision of technology transfer services. All of this activity was at smaller colleges with fewer than 20,000 students, and 80% of it was accounted for by organizations that describe themselves as "A college or university that is not a major research university." More than 62% of colleges with 3 or fewer employees in the technology transfer office have engaged in mergers or other forms of re-organization with other college offices or departments.

HOW TO USE BENCHMARKING DATA FROM:

Global Higher Education Technology Transfer Office Outreach & Marketing Activities

Global Higher Education Technology Transfer Office Outreach & Marketing Activities is derived from the popular 227-page report, *Global Higher Education Technology Transfer Office Benchmarks* (ISBN: 1-57440-167-X). This condensed incarnation focuses solely on internal and external marketing, public relations, and partnerships, presenting 204 benchmarking tables with over 1,000 comparative data sets. The best methods for making maximum use of the large number of different data sets for the academic years of 2009, 2010, and 2011 is to:

1. Take actual figures from your institution's normal financial and business reports and make general one to one comparisons:
    a. First with the MEAN numbers
    b. Secondly with the best category for your institution
    c. Thirdly examine the median and minimum and maximum ranges
    d. Finally analyze the unique factors for your institution
2. This information is useful for:
    a. Budget planning cycles and analysis
    b. Personnel and Salary Planning
    c. Internal marketing efforts
    d. External marketing efforts
    e. Legal activities and expenditures
    f. Three and five year planning projections and goals.

Participation: The participating colleges and universities included institutions from the United States, Europe and Asia.

Topics Covered: Each participating technology transfer office fully completed a detailed questionnaire on Staffing, Salaries, Budgets, Legal Services used Internally and Externally, Public Relations, Marketing Internally and Externally, Library Services Usage, and Partnering Activities.

Data was broken out by the following factors:

1. Location of the Global Higher Education Institution:
    USA, Other [outside the US] and All [Global]

2. Size of Higher Education's Enrollment by Number Of Students:
   Under 10,000; 10,000 to 19,999; 20,000 to 29,999, Over 30,000; Hospital
3. Type of the Global Higher Education Institution:
   Major Research University; College or University but not a Major Research University; and a specialized institution of higher education such as medical school or engineering school
4. Size of Technology Transfer Office by Number of Employees:
   Under 3; 4 to 7; 8 to 13; over 14
5. Technology Transfer Office by Types of Technology Concentration:
   Energy; Medical; Engineering and Informatics; and Mining-Materials-Agriculture

Users of the Benchmark information include: Technology Transfer Offices, Higher Education Schools such as medical, engineering, materials, energy, agriculture, computer science, sciences, and many others; University Departments of Provost, Grants and Contracts, Auditing, Finance, Budgeting, Public Relations, Library Services, and many others; Law Firms, Accounting Firms, Public Relations Firms, Venture Capital, Corporation Licensing and Technology Transfer Offices, Business Schools, Incubators, Local, State and Federal Economic Development Agencies and Technology Service Providers of all types.

Executive Summary of the Chapters

Chapter One: Internal and External Marketing and Technology Development

A. Invention reports are seen increasing by 5.2% in 2011.
B. Complete with 120 Tables of Internal and External Marketing

Chapter Two: Developments in Public Relations

A. Mean spending on website development in 2010 was $2,979.20.
B. Complete with 60 Tables of Public Outreach Expenses and Activities.

Chapter Three: Partnerships

A. Partnerships seen rising by 20% in 2011.
B. Complete with 24 Tables on Partnership Activities.

Chapter One: Internal and External Marketing and Technology Development

Question One: How many invention disclosure reports did the office receive from faculty in 2009?

Table 1.1: How many invention disclosure reports did the office receive from faculty in 2009?

	Mean	Median	Minimum	Maximum
Entire Sample	80.70	67.50	2.00	352.00

Table 1.2: How many invention disclosure reports did the office receive from faculty in 2009? Broken out by country.

Country	Mean	Median	Minimum	Maximum
USA	91.59	76.00	4.00	352.00
Other	50.75	35.00	2.00	150.00

Table 1.3: How many invention disclosure reports did the office receive from faculty in 2009? Broken out by enrollment.

Enrollment	Mean	Median	Minimum	Maximum
Under 10,000	65.38	61.00	2.00	148.00
10,000 to 19,999	62.10	33.50	4.00	352.00
20,000 to 29,999	104.14	91.00	65.00	179.00
30,000 and above	217.00	217.00	84.00	350.00
Hospital	38.00	27.00	12.00	75.00

Table 1.4: How many invention disclosure reports did the office receive from faculty in 2009? Broken out by technology area.

Technology Area	Mean	Median	Minimum	Maximum
Energy	33.67	8.00	2.00	91.00
Medical	103.79	75.00	12.00	352.00
Engineering and Informatics	32.20	30.00	4.00	84.00
Mining, Materials & Agriculture	62.33	70.00	40.00	77.00

Table 1.5: How many invention disclosure reports did the office receive from faculty in 2009? Broken out by what the technology transfer office serves.

Institution	Mean	Median	Minimum	Maximum
A specialized unit or college such as a medical school or engineering college	38.00	27.00	12.00	75.00
Major research university	96.83	77.00	4.00	352.00
A college or university that is not a major research university	20.00	19.00	2.00	40.00

Table 1.6: How many invention disclosure reports did the office receive from faculty in 2009? Broken out by the number of full time employees in the technology transfer office in 2009.

# of Employees	Mean	Median	Minimum	Maximum
3 or fewer	16.14	12.00	4.00	30.00
4-7	65.82	70.00	2.00	179.00
8-13	63.80	75.00	15.00	82.00
14 or more	180.71	148.00	35.00	352.00

Question Two: How many invention disclosure reports did the office receive from faculty in 2010

Table 1.7: How many invention disclosure reports did the office receive from faculty in 2010?

	Mean	Median	Minimum	Maximum
Entire Sample	77.65	54.00	3.00	355.00

Table 1.8: How many invention disclosure reports did the office receive from faculty in 2010? Broken out by country.

Country	Mean	Median	Minimum	Maximum
USA	85.96	67.00	3.00	355.00
Other	53.75	37.00	4.00	150.00

Table 1.9: How many invention disclosure reports did the office receive from faculty in 2010? Broken out by enrollment.

Enrollment	Mean	Median	Minimum	Maximum
Under 10,000	73.00	56.00	4.00	216.00
10,000 to 19,999	58.18	30.00	3.00	355.00
20,000 to 29,999	95.14	74.00	54.00	150.00
30,000 and above	190.50	190.50	91.00	290.00
Hospital	45.33	25.00	12.00	99.00

Table 1.10: How many invention disclosure reports did the office receive from faculty in 2010? Broken out by technology area.

Technology Area	Mean	Median	Minimum	Maximum
Energy	27.00	10.00	4.00	67.00
Medical	103.47	74.00	12.00	355.00
Engineering and Informatics	30.50	26.00	3.00	91.00
Mining, Materials & Agriculture	59.00	54.00	40.00	83.00

Table 1.11: How many invention disclosure reports did the office receive from faculty in 2010? Broken out by what the technology transfer office serves.

Institution	Mean	Median	Minimum	Maximum
A specialized unit or college such as a medical school or engineering college	45.33	25.00	12.00	99.00
Major research university	95.91	73.00	3.00	355.00
A college or university that is not a major research university	13.00	10.00	3.00	30.00

Table 1.12: How many invention disclosure reports did the office receive from faculty in 2010? Broken out by the number of full time employees in the technology transfer office in 2009.

# of Employees	Mean	Median	Minimum	Maximum
3 or fewer	15.38	15.00	3.00	30.00
4-7	59.55	54.00	4.00	148.00
8-13	70.60	73.00	25.00	99.00
14 or more	182.29	150.00	40.00	355.00

Question Three: How many invention disclosure reports are expected to be received by the office from faculty in 2011?

Table 1.13: How many invention disclosure reports are expected to be received by the office from faculty in 2011?

	Mean	Median	Minimum	Maximum
Entire Sample	81.43	65.00	3.00	360.00

Table 1.14: How many invention disclosure reports are expected to be received by the office from faculty in 2011? Broken out by country.

Country	Mean	Median	Minimum	Maximum
USA	91.27	75.00	12.00	360.00
Other	54.38	35.50	3.00	150.00

Table 1.15: How many invention disclosure reports are expected to be received by the office from faculty in 2011? Broken out by enrollment.

Enrollment	Mean	Median	Minimum	Maximum
Under 10,000	70.00	60.00	3.00	160.00
10,000 to 19,999	68.30	35.50	12.00	360.00
20,000 to 29,999	99.29	85.00	60.00	160.00
30,000 and above	192.50	192.50	85.00	300.00
Hospital	40.00	25.00	15.00	80.00

Table 1.16: How many invention disclosure reports are expected to be received by the office from faculty in 2011? Broken out by technology area.

Technology Area	Mean	Median	Minimum	Maximum
Energy	29.33	15.00	3.00	70.00
Medical	104.58	80.00	15.00	360.00
Engineering and Informatics	37.60	30.00	12.00	85.00
Mining, Materials & Agriculture	60.00	60.00	40.00	80.00

Table 1.17: How many invention disclosure reports are expected to be received by the office from faculty in 2011? Broken out by what the technology transfer office serves.

Institution	Mean	Median	Minimum	Maximum
A specialized unit or college such as a medical school or engineering college	40.00	25.00	15.00	80.00
Major research university	101.05	80.00	24.00	360.00
A college or university that is not a major research university	20.00	15.00	3.00	40.00

Table 1.18: How many invention disclosure reports are expected to be received by the office from faculty in 2011? Broken out by the number of full time employees in the technology transfer office in 2009.

# of Employees	Mean	Median	Minimum	Maximum
3 or fewer	20.86	24.00	12.00	30.00
4-7	64.45	60.00	3.00	160.00
8-13	68.60	80.00	26.00	87.00
14 or more	177.86	150.00	35.00	360.00

Question Four: What percentage of the invention disclosure reports was the office initially contacted by the faculty member in 2009 (as opposed to solicited by the office itself)?

Table 1.19: What percentage of the invention disclosure reports was the office initially contacted by the faculty member in 2009 (as opposed to solicited by the office itself)?

	Mean	Median	Minimum	Maximum
Entire Sample	69.00%	80.00%	2.00%	100.00%

Table 1.20: What percentage of the invention disclosure reports was the office initially contacted by the faculty member in 2009 (as opposed to solicited by the office itself)? Broken out by country.

Country	Mean	Median	Minimum	Maximum
USA	75.10%	80.00%	2.00%	100.00%
Other	53.00%	50.00%	6.00%	100.00%

Table 1.21: What percentage of the invention disclosure reports was the office initially contacted by the faculty member in 2009 (as opposed to solicited by the office itself)? Broken out by enrollment.

Enrollment	Mean	Median	Minimum	Maximum
Under 10,000	68.88%	77.50%	6.00%	100.00%
10,000 to 19,999	76.89%	90.00%	2.00%	100.00%
20,000 to 29,999	56.33%	55.00%	18.00%	85.00%
30,000 and above	71.67%	90.00%	30.00%	95.00%
Hospital	68.33%	75.00%	50.00%	80.00%

Table 1.22: What percentage of the invention disclosure reports was the office initially contacted by the faculty member in 2009 (as opposed to solicited by the office itself)? Broken out by technology area.

Technology Area	Mean	Median	Minimum	Maximum
Energy	48.67%	40.00%	6.00%	100.00%
Medical	69.37%	80.00%	18.00%	100.00%
Engineering and Informatics	76.40%	90.00%	2.00%	100.00%
Mining, Materials & Agriculture	77.50%	77.50%	70.00%	85.00%

Table 1.23: What percentage of the invention disclosure reports was the office initially contacted by the faculty member in 2009 (as opposed to solicited by the office itself)? Broken out by what the technology transfer office serves.

Institution	Mean	Median	Minimum	Maximum
A specialized unit or college such as a medical school or engineering college	68.33%	75.00%	50.00%	80.00%
Major research university	69.09%	80.00%	2.00%	100.00%
A college or university that is not a major research university	69.00%	85.00%	6.00%	100.00%

Table 1.24: What percentage of the invention disclosure reports was the office initially contacted by the faculty member in 2009 (as opposed to solicited by the office itself)? Broken out by the number of full time employees in the technology transfer office in 2009.

# of Employees	Mean	Median	Minimum	Maximum
3 or fewer	74.57%	90.00%	2.00%	100.00%
4-7	66.00%	80.00%	6.00%	100.00%
8-13	59.60%	75.00%	18.00%	80.00%
14 or more	75.83%	85.00%	50.00%	95.00%

Question Five: For what percentage of the invention disclosure reports was the office initially contacted by the faculty member in 2010 (as opposed to solicited by the office itself)?

Table 1.25: For what percentage of the invention disclosure reports was the office initially contacted by the faculty member in 2010 (as opposed to solicited by the office itself)?

	Mean	Median	Minimum	Maximum
Entire Sample	66.76%	80.00%	2.00%	100.00%

Table 1.26: For what percentage of the invention disclosure reports was the office initially contacted by the faculty member in 2010 (as opposed to solicited by the office itself)? Broken out by country.

Country	Mean	Median	Minimum	Maximum
USA	72.00%	80.00%	2.00%	100.00%
Other	53.00%	50.00%	6.00%	100.00%

Table 1.27: For what percentage of the invention disclosure reports was the office initially contacted by the faculty member in 2010 (as opposed to solicited by the office itself)? Broken out by enrollment.

Enrollment	Mean	Median	Minimum	Maximum
Under 10,000	68.88%	77.50%	6.00%	100.00%
10,000 to 19,999	68.00%	80.00%	2.00%	100.00%
20,000 to 29,999	57.17%	55.00%	18.00%	85.00%
30,000 and above	73.33%	90.00%	35.00%	95.00%
Hospital	70.00%	80.00%	50.00%	80.00%

Table 1.28: For what percentage of the invention disclosure reports was the office initially contacted by the faculty member in 2010 (as opposed to solicited by the office itself)? Broken out by technology area.

Technology Area	Mean	Median	Minimum	Maximum
Energy	42.00%	50.00%	6.00%	70.00%
Medical	69.63%	80.00%	18.00%	100.00%
Engineering and Informatics	66.40%	90.00%	2.00%	100.00%
Mining, Materials & Agriculture	77.50%	77.50%	70.00%	85.00%

Table 1.29: For what percentage of the invention disclosure reports was the office initially contacted by the faculty member in 2010 (as opposed to solicited by the office itself)? Broken out by what the technology transfer office serves.

Institution	Mean	Median	Minimum	Maximum
A specialized unit or college such as a medical school or engineering college	70.00%	80.00%	50.00%	80.00%
Major research university	67.27%	77.50%	2.00%	100.00%
A college or university that is not a major research university	61.50%	75.00%	6.00%	90.00%

Table 1.30: For what percentage of the invention disclosure reports was the office initially contacted by the faculty member in 2010 (as opposed to solicited by the office itself)? Broken out by the number of full time employees in the technology transfer office in 2009.

# of Employees	Mean	Median	Minimum	Maximum
3 or fewer	63.14%	70.00%	2.00%	100.00%
4-7	66.91%	80.00%	6.00%	100.00%
8-13	60.60%	75.00%	18.00%	80.00%
14 or more	75.83%	85.00%	50.00%	95.00%

Question Six: What percentage of invention disclosure reports were proffered by a faculty member in 2011 (as opposed to solicited by the office itself)?

Table 1.31: What percentage of the invention disclosure reports were proffered by the faculty member in 2011 (as opposed to solicited by the office itself)?

	Mean	Median	Minimum	Maximum
Entire Sample	67.43%	75.00%	4.00%	100.00%

Table 1.32: What percentage of the invention disclosure reports were proffered by the faculty member in 2011 (as opposed to solicited by the office itself)? Broken out by country.

Country	Mean	Median	Minimum	Maximum
USA	72.95%	77.50%	4.00%	100.00%
Other	53.63%	50.00%	6.00%	100.00%

Table 1.33: What percentage of the invention disclosure reports were proffered by the faculty member in 2011 (as opposed to solicited by the office itself)? Broken out by enrollment.

Enrollment	Mean	Median	Minimum	Maximum
Under 10,000	68.88%	77.50%	6.00%	100.00%
10,000 to 19,999	70.50%	80.00%	4.00%	100.00%
20,000 to 29,999	57.17%	62.50%	18.00%	75.00%
30,000 and above	75.00%	90.00%	40.00%	95.00%
Hospital	68.33%	75.00%	50.00%	80.00%

Table 1.34: What percentage of the invention disclosure reports were proffered by the faculty member in 2011 (as opposed to solicited by the office itself)? Broken out by technology area.

Technology Area	Mean	Median	Minimum	Maximum
Energy	30.50%	30.50%	6.00%	55.00%
Medical	69.89%	75.00%	18.00%	100.00%
Engineering and Informatics	70.80%	85.00%	4.00%	100.00%
Mining, Materials & Agriculture	72.50%	72.50%	70.00%	75.00%

Table 1.35: What percentage of the invention disclosure reports were proffered by the faculty member in 2011 (as opposed to solicited by the office itself)? Broken out by what the technology transfer office serves.

Institution	Mean	Median	Minimum	Maximum
A specialized unit or college such as a medical school or engineering college	68.33%	75.00%	50.00%	80.00%
Major research university	68.73%	75.00%	4.00%	100.00%
A college or university that is not a major research university	57.00%	80.00%	6.00%	85.00%

Table 1.36: What percentage of the invention disclosure reports were proffered by the faculty member in 2011 (as opposed to solicited by the office itself)? Broken out by the number of full time employees in the technology transfer office in 2009.

# of Employees	Mean	Median	Minimum	Maximum
3 or fewer	65.67%	77.50%	4.00%	100.00%
4-7	67.36%	75.00%	6.00%	100.00%
8-13	59.60%	75.00%	18.00%	80.00%
14 or more	75.83%	85.00%	50.00%	95.00%

Question Seven: What percentage resulted from strong encouragement by the technology transfer office in 2009?

Table 1.37: What percentage resulted from strong encouragement by the technology transfer office in 2009?

	Mean	Median	Minimum	Maximum
Entire Sample	25.04%	20.00%	0.00%	100.00%

Table 1.38: What percentage resulted from strong encouragement by the technology transfer office in 2009? Broken out by country.

Country	Mean	Median	Minimum	Maximum
USA	20.26%	15.00%	0.00%	100.00%
Other	36.38%	30.00%	0.00%	82.00%

Table 1.39: What percentage resulted from strong encouragement by the technology transfer office in 2009? Broken out by enrollment.

Enrollment	Mean	Median	Minimum	Maximum
Under 10,000	14.86%	10.00%	0.00%	40.00%
10,000 to 19,999	25.56%	10.00%	0.00%	100.00%
20,000 to 29,999	40.33%	25.00%	15.00%	82.00%
30,000 and above	10.00%	5.00%	5.00%	20.00%
Hospital	35.00%	35.00%	20.00%	50.00%

Table 1.40: What percentage resulted from strong encouragement by the technology transfer office in 2009? Broken out by technology area.

Technology Area	Mean	Median	Minimum	Maximum
Energy	13.00%	9.00%	0.00%	30.00%
Medical	29.82%	20.00%	0.00%	82.00%
Engineering and Informatics	23.00%	5.00%	0.00%	100.00%
Mining, Materials & Agriculture	7.50%	7.50%	0.00%	15.00%

Table 1.41: What percentage resulted from strong encouragement by the technology transfer office in 2009? Broken out by what the technology transfer office serves.

Institution	Mean	Median	Minimum	Maximum
A specialized unit or college such as a medical school or engineering college	35.00%	35.00%	20.00%	50.00%
Major research university	27.00%	20.00%	0.00%	100.00%
A college or university that is not a major research university	9.75%	9.50%	0.00%	20.00%

Table 1.42: What percentage resulted from strong encouragement by the technology transfer office in 2009? Broken out by the number of full time employees in the technology transfer office in 2009.

# of Employees	Mean	Median	Minimum	Maximum
3 or fewer	25.71%	10.00%	0.00%	100.00%
4-7	13.09%	15.00%	0.00%	30.00%
8-13	49.00%	40.00%	25.00%	82.00%
14 or more	34.17%	30.00%	5.00%	80.00%

Question Eight: What percentage resulted from strong encouragement by the technology transfer office in 2010?

Table 1.43: What percentage resulted from strong encouragement by the technology transfer office in 2010?

	Mean	Median	Minimum	Maximum
Entire Sample	24.11%	20.00%	0.00%	82.00%

Table 1.44: What percentage resulted from strong encouragement by the technology transfer office in 2010? Broken out by country.

Country	Mean	Median	Minimum	Maximum
USA	18.95%	20.00%	0.00%	50.00%
Other	36.38%	30.00%	0.00%	82.00%

Table 1.45: What percentage resulted from strong encouragement by the technology transfer office in 2010? Broken out by enrollment.

Enrollment	Mean	Median	Minimum	Maximum
Under 10,000	14.86%	10.00%	0.00%	40.00%
10,000 to 19,999	22.22%	20.00%	0.00%	50.00%
20,000 to 29,999	41.17%	25.00%	15.00%	82.00%
30,000 and above	10.00%	5.00%	5.00%	20.00%
Hospital	35.00%	35.00%	20.00%	50.00%

Table 1.46: What percentage resulted from strong encouragement by the technology transfer office in 2010? Broken out by technology area.

Technology Area	Mean	Median	Minimum	Maximum
Energy	19.67%	20.00%	9.00%	30.00%
Medical	30.12%	20.00%	0.00%	82.00%
Engineering and Informatics	13.00%	5.00%	0.00%	50.00%
Mining, Materials & Agriculture	7.50%	7.50%	0.00%	15.00%

Table 1.47: What percentage resulted from strong encouragement by the technology transfer office in 2010? Broken out by what the technology transfer office serves.

Institution	Mean	Median	Minimum	Maximum
A specialized unit or college such as a medical school or engineering college	35.00%	35.00%	20.00%	50.00%
Major research university	24.86%	20.00%	0.00%	82.00%
A college or university that is not a major research university	14.75%	15.00%	9.00%	20.00%

Table 1.48: What percentage resulted from strong encouragement by the technology transfer office in 2010? Broken out by the number of full time employees in the technology transfer office in 2009.

# of Employees	Mean	Median	Minimum	Maximum
3 or fewer	21.43%	20.00%	0.00%	50.00%
4-7	13.55%	15.00%	0.00%	30.00%
8-13	49.00%	40.00%	25.00%	82.00%
14 or more	34.17%	30.00%	5.00%	80.00%

Question Nine: What percentage can result from strong encouragement by the technology transfer office in 2011?

Table 1.49: What percentage can result from strong encouragement by the technology transfer office in 2011?

	Mean	Median	Minimum	Maximum
Entire Sample	28.31%	20.00%	0.00%	82.00%

Table 1.50: What percentage can result from strong encouragement by the technology transfer office in 2011? Broken out by country.

Country	Mean	Median	Minimum	Maximum
USA	23.89%	20.00%	0.00%	75.00%
Other	38.25%	35.00%	0.00%	82.00%

Table 1.51: What percentage can result from strong encouragement by the technology transfer office in 2011? Broken out by enrollment.

Enrollment	Mean	Median	Minimum	Maximum
Under 10,000	16.29%	10.00%	0.00%	50.00%
10,000 to 19,999	32.50%	30.00%	0.00%	75.00%
20,000 to 29,999	43.67%	27.50%	20.00%	82.00%
30,000 and above	10.00%	5.00%	5.00%	20.00%
Hospital	35.00%	35.00%	20.00%	50.00%

Table 1.52: What percentage can result from strong encouragement by the technology transfer office in 2011? Broken out by technology area.

Technology Area	Mean	Median	Minimum	Maximum
Energy	19.50%	19.50%	9.00%	30.00%
Medical	31.00%	20.00%	0.00%	82.00%
Engineering and Informatics	29.00%	15.00%	0.00%	75.00%
Mining, Materials & Agriculture	12.50%	12.50%	0.00%	25.00%

Table 1.53: What percentage can result from strong encouragement by the technology transfer office in 2011? Broken out by what the technology transfer office serves.

Institution	Mean	Median	Minimum	Maximum
A specialized unit or college such as a medical school or engineering college	35.00%	35.00%	20.00%	50.00%
Major research university	29.62%	25.00%	0.00%	82.00%
A college or university that is not a major research university	14.67%	15.00%	9.00%	20.00%

Table 1.54: What percentage can result from strong encouragement by the technology transfer office in 2011? Broken out by the number of full time employees in the technology transfer office in 2009.

# of Employees	Mean	Median	Minimum	Maximum
3 or fewer	35.00%	35.00%	0.00%	75.00%
4-7	14.91%	20.00%	0.00%	30.00%
8-13	52.33%	50.00%	25.00%	82.00%
14 or more	34.17%	30.00%	5.00%	80.00%

Question Ten: Are any employees only responsible for internal marketing to faculty? If so, how many in 2009?

Table 1.55: Are any employees only responsible for internal marketing to faculty? If so, how many in 2009?

	Mean	Median	Minimum	Maximum
Entire Sample	0.30	0.00	0.00	2.00

Table 1.56: Are any employees only responsible for internal marketing to faculty? If so, how many in 2009? Broken out by country.

Country	Mean	Median	Minimum	Maximum
USA	0.08	0.00	0.00	1.00
Other	0.81	0.50	0.00	2.00

Table 1.57: Are any employees only responsible for internal marketing to faculty? If so, how many in 2009? Broken out by enrollment.

Enrollment	Mean	Median	Minimum	Maximum
Under 10,000	0.64	0.50	0.00	2.00
10,000 to 19,999	0.19	0.00	0.00	1.50
20,000 to 29,999	0.33	0.00	0.00	2.00
30,000 and above	0.00	0.00	0.00	0.00
Hospital	0.00	0.00	0.00	0.00

Table 1.58: Are any employees only responsible for internal marketing to faculty? If so, how many in 2009? Broken out by technology area.

Technology Area	Mean	Median	Minimum	Maximum
Energy	0.33	0.00	0.00	1.00
Medical	0.41	0.00	0.00	2.00
Engineering and Informatics	0.00	0.00	0.00	0.00
Mining, Materials & Agriculture	0.00	0.00	0.00	0.00

Table 1.59: Are any employees only responsible for internal marketing to faculty? If so, how many in 2009? Broken out by what the technology transfer office serves.

Institution	Mean	Median	Minimum	Maximum
A specialized unit or college such as a medical school or engineering college	0.00	0.00	0.00	0.00
Major research university	0.33	0.00	0.00	2.00
A college or university that is not a major research university	0.33	0.00	0.00	1.00

Table 1.60: Are any employees only responsible for internal marketing to faculty? If so, how many in 2009? Broken out by the number of full time employees in the technology transfer office in 2009.

# of Employees	Mean	Median	Minimum	Maximum
3 or fewer	0.00	0.00	0.00	0.00
4-7	0.10	0.00	0.00	1.00
8-13	0.75	0.50	0.00	2.00
14 or more	0.67	0.25	0.00	2.00

Question Eleven: Are any employees only responsible for internal marketing to faculty? If so, how many in 2010?

Table 1.61: Are any employees only responsible for internal marketing to faculty? If so, how many in 2010?

	Mean	Median	Minimum	Maximum
Entire Sample	0.33	0.00	0.00	2.00

Table 1.62: Are any employees only responsible for internal marketing to faculty? If so, how many in 2010? Broken out by country.

Country	Mean	Median	Minimum	Maximum
USA	0.13	0.00	0.00	1.00
Other	0.81	0.50	0.00	2.00

Table 1.63: Are any employees only responsible for internal marketing to faculty? If so, how many in 2010? Broken out by enrollment.

Enrollment	Mean	Median	Minimum	Maximum
Under 10,000	0.64	0.50	0.00	2.00
10,000 to 19,999	0.31	0.00	0.00	1.50
20,000 to 29,999	0.33	0.00	0.00	2.00
30,000 and above	0.00	0.00	0.00	0.00
Hospital	0.00	0.00	0.00	0.00

Table 1.64: Are any employees only responsible for internal marketing to faculty? If so, how many in 2010? Broken out by technology area.

Technology Area	Mean	Median	Minimum	Maximum
Energy	0.33	0.00	0.00	1.00
Medical	0.41	0.00	0.00	2.00
Engineering and Informatics	0.20	0.00	0.00	1.00
Mining, Materials & Agriculture	0.00	0.00	0.00	0.00

Table 1.65: Are any employees only responsible for internal marketing to faculty? If so, how many in 2010? Broken out by what the technology transfer office serves.

Institution	Mean	Median	Minimum	Maximum
A specialized unit or college such as a medical school or engineering college	0.00	0.00	0.00	0.00
Major research university	0.38	0.00	0.00	2.00
A college or university that is not a major research university	0.33	0.00	0.00	1.00

Table 1.66: Are any employees only responsible for internal marketing to faculty? If so, how many in 2010? Broken out by the number of full time employees in the technology transfer office in 2009.

# of Employees	Mean	Median	Minimum	Maximum
3 or fewer	0.14	0.00	0.00	1.00
4-7	0.10	0.00	0.00	1.00
8-13	0.75	0.50	0.00	2.00
14 or more	0.67	0.25	0.00	2.00

Question Twelve: How many employees are planned to be only responsible for internal marketing to faculty in 2011?

Table 1.67: How many employees are planned to be only responsible for internal marketing to faculty in 2011?

	Mean	Median	Minimum	Maximum
Entire Sample	0.44	0.00	0.00	3.00

Table 1.68: How many employees are planned to be only responsible for internal marketing to faculty in 2011? Broken out by country.

Country	Mean	Median	Minimum	Maximum
USA	0.22	0.00	0.00	2.00
Other	0.94	0.50	0.00	3.00

Table 1.69: How many employees are planned to be only responsible for internal marketing to faculty in 2011? Broken out by enrollment.

Enrollment	Mean	Median	Minimum	Maximum
Under 10,000	0.71	1.00	0.00	2.00
10,000 to 19,999	0.50	0.00	0.00	2.00
20,000 to 29,999	0.50	0.00	0.00	3.00
30,000 and above	0.00	0.00	0.00	0.00
Hospital	0.00	0.00	0.00	0.00

Table 1.70: How many employees are planned to be only responsible for internal marketing to faculty in 2011? Broken out by technology area.

Technology Area	Mean	Median	Minimum	Maximum
Energy	0.50	0.50	0.00	1.00
Medical	0.50	0.00	0.00	3.00
Engineering and Informatics	0.40	0.00	0.00	2.00
Mining, Materials & Agriculture	0.00	0.00	0.00	0.00

Table 1.71: How many employees are planned to be only responsible for internal marketing to faculty in 2011? Broken out by what the technology transfer office serves.

Institution	Mean	Median	Minimum	Maximum
A specialized unit or college such as a medical school or engineering college	0.00	0.00	0.00	0.00
Major research university	0.50	0.00	0.00	3.00
A college or university that is not a major research university	0.50	0.50	0.00	1.00

Table 1.72: How many employees are planned to be only responsible for internal marketing to faculty in 2011? Broken out by the number of full time employees in the technology transfer office in 2009.

# of Employees	Mean	Median	Minimum	Maximum
3 or fewer	0.33	0.00	0.00	2.00
4-7	0.10	0.00	0.00	1.00
8-13	0.75	0.50	0.00	2.00
14 or more	0.92	0.50	0.00	3.00

Question Thirteen: Are any employees responsible for only external marketing to Industry? How many in 2009?

Table 1.73: Are any employees responsible for only external marketing to Industry? How many in 2009?

	Mean	Median	Minimum	Maximum
Entire Sample	0.81	0.00	0.00	8.00

Table 1.74: Are any employees responsible for only external marketing to Industry? How many in 2009? Broken out by country.

Country	Mean	Median	Minimum	Maximum
USA	0.82	0.00	0.00	8.00
Other	0.81	0.00	0.00	3.00

Table 1.75: Are any employees responsible for only external marketing to Industry? How many in 2009? Broken out by enrollment.

Enrollment	Mean	Median	Minimum	Maximum
Under 10,000	1.43	1.00	0.00	3.00
10,000 to 19,999	0.44	0.00	0.00	1.50
20,000 to 29,999	0.00	0.00	0.00	0.00
30,000 and above	2.67	0.00	0.00	8.00
Hospital	0.17	0.00	0.00	0.50

Table 1.76: Are any employees responsible for only external marketing to Industry? How many in 2009? Broken out by technology area.

Technology Area	Mean	Median	Minimum	Maximum
Energy	1.00	1.00	0.00	2.00
Medical	1.06	0.00	0.00	8.00
Engineering and Informatics	0.20	0.00	0.00	1.00
Mining, Materials & Agriculture	0.00	0.00	0.00	0.00

Table 1.77: Are any employees responsible for only external marketing to Industry? How many in 2009? Broken out by what the technology transfer office serves.

Institution	Mean	Median	Minimum	Maximum
A specialized unit or college such as a medical school or engineering college	0.17	0.00	0.00	0.50
Major research university	0.88	0.00	0.00	8.00
A college or university that is not a major research university	1.00	1.00	0.00	2.00

Table 1.78: Are any employees responsible for only external marketing to Industry? How many in 2009? Broken out by the number of full time employees in the technology transfer office in 2009.

# of Employees	Mean	Median	Minimum	Maximum
3 or fewer	0.57	0.00	0.00	2.00
4-7	0.20	0.00	0.00	2.00
8-13	1.13	0.75	0.00	3.00
14 or more	1.92	1.00	0.00	8.00

Question Fourteen: Are any employees responsible for only external marketing to Industry? How many in 2010?

Table 1.79: Are any employees responsible for only external marketing to Industry? How many in 2010?

	Mean	Median	Minimum	Maximum
Entire Sample	0.93	0.00	0.00	8.00

Table 1.80: Are any employees responsible for only external marketing to Industry? How many in 2010? Broken out by country.

Country	Mean	Median	Minimum	Maximum
USA	0.97	0.50	0.00	8.00
Other	0.81	0.00	0.00	3.00

Table 1.81: Are any employees responsible for only external marketing to Industry? How many in 2010? Broken out by enrollment.

Enrollment	Mean	Median	Minimum	Maximum
Under 10,000	1.43	1.00	0.00	3.00
10,000 to 19,999	0.69	0.50	0.00	2.00
20,000 to 29,999	0.00	0.00	0.00	0.00
30,000 and above	3.00	1.00	0.00	8.00
Hospital	0.17	0.00	0.00	0.50

Table 1.82: Are any employees responsible for only external marketing to Industry? How many in 2010? Broken out by technology area.

Technology Area	Mean	Median	Minimum	Maximum
Energy	1.33	2.00	0.00	2.00
Medical	1.18	1.00	0.00	8.00
Engineering and Informatics	0.20	0.00	0.00	1.00
Mining, Materials & Agriculture	0.00	0.00	0.00	0.00

Table 1.83: Are any employees responsible for only external marketing to Industry? How many in 2010? Broken out by what the technology transfer office serves.

Institution	Mean	Median	Minimum	Maximum
A specialized unit or college such as a medical school or engineering college	0.17	0.00	0.00	0.50
Major research university	0.98	0.00	0.00	8.00
A college or university that is not a major research university	1.33	2.00	0.00	2.00

Table 1.84: Are any employees responsible for only external marketing to Industry? How many in 2010? Broken out by the number of full time employees in the technology transfer office in 2009.

# of Employees	Mean	Median	Minimum	Maximum
3 or fewer	0.71	0.00	0.00	2.00
4-7	0.30	0.00	0.00	2.00
8-13	1.13	0.75	0.00	3.00
14 or more	2.08	1.00	0.00	8.00

Question Fifteen: Are any employees responsible for only external marketing to Industry? How many in 2011?

Table 1.85: Are any employees responsible for only external marketing to Industry? How many in 2011?

	Mean	Median	Minimum	Maximum
Entire Sample	1.02	0.50	0.00	8.00

Table 1.86: Are any employees responsible for only external marketing to Industry? How many in 2011? Broken out by country.

Country	Mean	Median	Minimum	Maximum
USA	1.11	0.60	0.00	8.00
Other	0.81	0.00	0.00	3.00

Table 1.87: Are any employees responsible for only external marketing to Industry? How many in 2011? Broken out by enrollment.

Enrollment	Mean	Median	Minimum	Maximum
Under 10,000	1.43	1.00	0.00	3.00
10,000 to 19,999	0.94	0.50	0.00	3.00
20,000 to 29,999	0.10	0.00	0.00	0.60
30,000 and above	3.00	1.00	0.00	8.00
Hospital	0.17	0.00	0.00	0.50

Table 1.88: Are any employees responsible for only external marketing to Industry? How many in 2011? Broken out by technology area.

Technology Area	Mean	Median	Minimum	Maximum
Energy	1.67	2.00	0.00	3.00
Medical	1.24	1.00	0.00	8.00
Engineering and Informatics	0.20	0.00	0.00	1.00
Mining, Materials & Agriculture	0.30	0.30	0.00	0.60

Table 1.89: Are any employees responsible for only external marketing to Industry? How many in 2011? Broken out by what the technology transfer office serves.

Institution	Mean	Median	Minimum	Maximum
A specialized unit or college such as a medical school or engineering college	0.17	0.00	0.00	0.50
Major research university	1.05	0.60	0.00	8.00
A college or university that is not a major research university	1.67	2.00	0.00	3.00

Table 1.90: Are any employees responsible for only external marketing to Industry? How many in 2011? Broken out by the number of full time employees in the technology transfer office in 2009.

# of Employees	Mean	Median	Minimum	Maximum
3 or fewer	0.86	0.00	0.00	3.00
4-7	0.36	0.00	0.00	2.00
8-13	1.13	0.75	0.00	3.00
14 or more	2.25	1.25	0.00	8.00

Question Sixteen: How many professionals were dedicated to spin out companies in 2009?

Table 1.91: How many professionals were dedicated to spin out companies in 2009?

	Mean	Median	Minimum	Maximum
Entire Sample	0.54	0.00	0.00	3.00

Table 1.92: How many professionals were dedicated to spin out companies in 2009? Broken out by country.

Country	Mean	Median	Minimum	Maximum
USA	0.43	0.00	0.00	2.00
Other	0.81	0.00	0.00	3.00

Table 1.93: How many professionals were dedicated to spin out companies in 2009? Broken out by enrollment.

Enrollment	Mean	Median	Minimum	Maximum
Under 10,000	0.71	0.50	0.00	3.00
10,000 to 19,999	0.63	0.25	0.00	2.00
20,000 to 29,999	0.29	0.00	0.00	2.00
30,000 and above	1.00	1.00	0.00	2.00
Hospital	0.00	0.00	0.00	0.00

Table 1.94: How many professionals were dedicated to spin out companies in 2009? Broken out by technology area.

Technology Area	Mean	Median	Minimum	Maximum
Energy	1.00	0.00	0.00	3.00
Medical	0.42	0.00	0.00	2.00
Engineering and Informatics	0.80	1.00	0.00	2.00
Mining, Materials & Agriculture	0.25	0.25	0.00	0.50

Table 1.95: How many professionals were dedicated to spin out companies in 2009? Broken out by what the technology transfer office serves.

Institution	Mean	Median	Minimum	Maximum
A specialized unit or college such as a medical school or engineering college	0.00	0.00	0.00	0.00
Major research university	0.55	0.00	0.00	2.00
A college or university that is not a major research university	1.00	0.00	0.00	3.00

Table 1.96: How many professionals were dedicated to spin out companies in 2009? Broken out by the number of full time employees in the technology transfer office in 2009.

# of Employees	Mean	Median	Minimum	Maximum
3 or fewer	0.29	0.00	0.00	1.00
4-7	0.65	0.00	0.00	3.00
8-13	0.13	0.00	0.00	0.50
14 or more	0.86	0.50	0.00	2.00

Question Seventeen: How many professionals were dedicated to spin out companies in 2010?

Table 1.97: How many professionals were dedicated to spin out companies in 2010?

	Mean	Median	Minimum	Maximum
Entire Sample	0.55	0.00	0.00	3.00

Table 1.98: How many professionals were dedicated to spin out companies in 2010? Broken out by country.

Country	Mean	Median	Minimum	Maximum
USA	0.45	0.00	0.00	3.00
Other	0.81	0.00	0.00	3.00

Table 1.99: How many professionals were dedicated to spin out companies in 2010? Broken out by enrollment.

Enrollment	Mean	Median	Minimum	Maximum
Under 10,000	0.86	0.50	0.00	3.00
10,000 to 19,999	0.69	0.50	0.00	2.00
20,000 to 29,999	0.14	0.00	0.00	1.00
30,000 and above	1.00	0.00	0.00	3.00
Hospital	0.00	0.00	0.00	0.00

Table 1.100: How many professionals were dedicated to spin out companies in 2010? Broken out by technology area.

Technology Area	Mean	Median	Minimum	Maximum
Energy	1.00	0.00	0.00	3.00
Medical	0.50	0.00	0.00	3.00
Engineering and Informatics	0.60	0.00	0.00	2.00
Mining, Materials & Agriculture	0.25	0.25	0.00	0.50

Table 1.101: How many professionals were dedicated to spin out companies in 2010? Broken out by what the technology transfer office serves.

Institution	Mean	Median	Minimum	Maximum
A specialized unit or college such as a medical school or engineering college	0.00	0.00	0.00	0.00
Major research university	0.57	0.00	0.00	3.00
A college or university that is not a major research university	1.00	0.00	0.00	3.00

Table 1.102: How many professionals were dedicated to spin out companies in 2010? Broken out by the number of full time employees in the technology transfer office in 2009.

# of Employees	Mean	Median	Minimum	Maximum
3 or fewer	0.43	0.00	0.00	2.00
4-7	0.55	0.00	0.00	3.00
8-13	0.13	0.00	0.00	0.50
14 or more	0.93	1.00	0.00	3.00

Question Eighteen: How many professionals are dedicated to spin out companies in 2011?

Table 1.103: How many professionals are dedicated to spin out companies in 2011?

	Mean	Median	Minimum	Maximum
Entire Sample	0.57	0.00	0.00	3.00

Table 1.104: How many professionals are dedicated to spin out companies in 2011? Broken out by country.

Country	Mean	Median	Minimum	Maximum
USA	0.47	0.00	0.00	3.00
Other	0.81	0.00	0.00	3.00

Table 1.105: How many professionals are dedicated to spin out companies in 2011? Broken out by enrollment.

Enrollment	Mean	Median	Minimum	Maximum
Under 10,000	0.86	0.50	0.00	3.00
10,000 to 19,999	0.69	0.50	0.00	2.00
20,000 to 29,999	0.20	0.00	0.00	1.00
30,000 and above	1.00	0.00	0.00	3.00
Hospital	0.00	0.00	0.00	0.00

Table 1.106: How many professionals are dedicated to spin out companies in 2011? Broken out by technology area.

Technology Area	Mean	Median	Minimum	Maximum
Energy	1.00	0.00	0.00	3.00
Medical	0.50	0.00	0.00	3.00
Engineering and Informatics	0.60	0.00	0.00	2.00
Mining, Materials & Agriculture	0.45	0.45	0.40	0.50

Table 1.107: How many professionals are dedicated to spin out companies in 2011? Broken out by what the technology transfer office serves.

Institution	Mean	Median	Minimum	Maximum
A specialized unit or college such as a medical school or engineering college	0.00	0.00	0.00	0.00
Major research university	0.59	0.00	0.00	3.00
A college or university that is not a major research university	1.00	0.00	0.00	3.00

Table 1.108: How many professionals are dedicated to spin out companies in 2011? Broken out by the number of full time employees in the technology transfer office in 2009.

# of Employees	Mean	Median	Minimum	Maximum
3 or fewer	0.43	0.00	0.00	2.00
4-7	0.59	0.00	0.00	3.00
8-13	0.13	0.00	0.00	0.50
14 or more	0.93	1.00	0.00	3.00

Question Nineteen: Does the office use an outside public relations firm?

Table 1.109: Does the office use an outside public relations firm?

	Yes	No
Entire Sample	6.45%	93.55%

Table 1.110: Does the office use an outside public relations firm? Broken out by country.

Country	Yes	No
USA	0.00%	100.00%
Other	25.00%	75.00%

Table 1.111: Does the office use an outside public relations firm? Broken out by enrollment.

Enrollment	Yes	No
Under 10,000	12.50%	87.50%
10,000 to 19,999	11.11%	88.89%
20,000 to 29,999	0.00%	100.00%
30,000 and above	0.00%	100.00%
Hospital	0.00%	100.00%

Table 1.112: Does the office use an outside public relations firm? Broken out by technology area.

Technology Area	Yes	No
Energy	25.00%	75.00%
Medical	0.00%	100.00%
Engineering and Informatics	25.00%	75.00%
Mining, Materials & Agriculture	0.00%	100.00%

Table 1.113: Does the office use an outside public relations firm? Broken out by what the technology transfer office serves.

Institution	Yes	No
A specialized unit or college such as a medical school or engineering college	0.00%	100.00%
Major research university	4.17%	95.83%
A college or university that is not a major research university	25.00%	75.00%

Table 1.114: Does the office use an outside public relations firm? Broken out by the number of full time employees in the technology transfer office in 2009.

# of Employees	Yes	No
3 or fewer	0.00%	100.00%
4-7	16.67%	83.33%
8-13	0.00%	100.00%
14 or more	0.00%	100.00%

Question Twenty: Does the office use the university public relations department?

Table 1.115: Does the office use the university public relations department?

	Yes	No
Entire Sample	84.85%	15.15%

Table 1.116: Does the office use the university public relations department? Broken out by country.

Country	Yes	No
USA	84.00%	16.00%
Other	87.50%	12.50%

Table 1.117: Does the office use the university public relations department? Broken out by enrollment.

Enrollment	Yes	No
Under 10,000	100.00%	0.00%
10,000 to 19,999	90.91%	9.09%
20,000 to 29,999	85.71%	14.29%
30,000 and above	33.33%	66.67%
Hospital	75.00%	25.00%

Table 1.118: Does the office use the university public relations department? Broken out by technology area.

Technology Area	Yes	No
Energy	50.00%	50.00%
Medical	90.00%	10.00%
Engineering and Informatics	83.33%	16.67%
Mining, Materials & Agriculture	100.00%	0.00%

Table 1.119: Does the office use the university public relations department? Broken out by what the technology transfer office serves.

Institution	Yes	No
A specialized unit or college such as a medical school or engineering college	100.00%	0.00%
Major research university	84.00%	16.00%
A college or university that is not a major research university	80.00%	20.00%

Table 1.120: Does the office use the university public relations department? Broken out by the number of full time employees in the technology transfer office in 2009.

# of Employees	Yes	No
3 or fewer	77.78%	22.22%
4-7	91.67%	8.33%
8-13	100.00%	0.00%
14 or more	71.43%	28.57%

Chapter Two: Developments in Public Relations

Question Twenty-one: Describe developments in your public relations efforts. How often do you send out releases? To whom do you send them? What results have you achieved and what do you recommend?

1. 6 press mailing lists, radio, TV.
2. Quarterly post cards; a few news releases; perhaps 6/year. Yes we recommend.
3. 1-2 / yr
4. Periodic press releases, usually in coordination with licensees, to announce new licenses.
5. Periodically.
6. Occasionally Boston Business Journal, Xconomy, BioWorld, Boston Globe some stories.
7. On an ad hoc basis.
8. Once or twice a year.
9. 1-2 times per year sent largely to our internal audience of employees, alumnae and Board of Trustees. Results are focused on awareness of IP issues and successes.
10. About 1 per week globally, to media and contact list – excellent tool for attracting partners and investors.
11. Several times a year.
12. Continuously.
13. Approx 6-8 annually.
14. 3-4 times a year to local papers. I would recommend more target releases on deals done to show success.
15. 5
16. When licenses are signed.
17. Periodically. Targeted to audiences based on the nature of the releases. We have received inquiries based on releases.
18. We send out releases infrequently but we do send them out when they are needed to generate commercial interest or to show past successes. These can be for internal intranet display or as wide as newspaper/online news release.

19. As and when we have noteworthy news. Press releases are sent to publications that suite the story. We have had mixed results depending on the story and its topicality. We are very much under resourced and do the best we can.
20. University Advancement sends out release almost daily.
21. Depends on what is meant here. We review releases by the departments/colleges and we also send our own out.
22. 2/year
23. 2x per year posted on TT website; need more aggressive marketing to reach intended customer base.
24. 1/month; broad distribution decided by University with supplemental suggestions by us (e.g., trade journals or societies relevant to a technology); excellent responses; recommend adding.
25. Determined case-by-case, targeted to potential licensees only.
26. Not often.
27. Tech transfer office does not issue press releases itself.
28. No.

Question Twenty-two: How often do you send out releases? To whom do you send them? What results have you achieved and what do you recommend?

1. University PR.
2. No money.
3. We work with University PR.
4. University office.
5. University PR.
6. Depends upon relationship with University PR. In our case, we don't get a great deal of traction within the University media relationship. If we had the budget I would use an outside firm.
7. In-house PR. Depends on your goals.
8. University PR.
9. We get assistance from the University Communications department.
10. Only use internal PR since we have no budget for this. No recommendation.
11. No.
12. University PR.
13. U. PR only no funding for outside use.
14. University PR is used and recommended for legal/exposure reasons.
15. No – develop in house capabilities because outside firms and university PR departments always do not understand the technology transfer field
16. 2
17. Our University communications department focuses on internal communications (i.e., intranet) but uses an external PR firm to get anything out in the wider media. This works in practice and allows us to keep an effective Communications department while utilizing the knowledge of the external PR firm.
18. We don't have the resource to go outside, our hard pressed marketing department provides us with some support for press releases. The University's prime function is to get bums on seats.
19. University PR/Marketing only.
20. We work with dept/college communications, university communications and we have our own dedicated marketing person.
21. Required to use university PR.

22. Use only University PR (Office of Research); University PR is excellent & no cost to us; recommend working with university PR first and using PR firm to supplement, if necessary.
23. University only.

Question Twenty-three: Do you send electronic releases?

Table 2.1: Do you send electronic releases?

	Yes	No
Entire Sample	75.68%	24.32%

Table 2.2: Do you send electronic releases? Broken out by country.

Country	Yes	No
USA	75.00%	25.00%
Other	77.78%	22.22%

Table 2.3: Do you send electronic releases? Broken out by enrollment.

Enrollment	Yes	No
Under 10,000	87.50%	12.50%
10,000 to 19,999	69.23%	30.77%
20,000 to 29,999	87.50%	12.50%
30,000 and above	75.00%	25.00%
Hospital	50.00%	50.00%

Table 2.4: Do you send electronic releases? Broken out by technology area.

Technology Area	Yes	No
Energy	25.00%	75.00%
Medical	82.61%	17.39%
Engineering and Informatics	71.43%	28.57%
Mining, Materials & Agriculture	100.00%	0.00%

Table 2.5: Do you send electronic releases? Broken out by what the technology transfer office serves.

Institution	Yes	No
A specialized unit or college such as a medical school or engineering college	50.00%	50.00%
Major research university	88.89%	11.11%
A college or university that is not a major research university	33.33%	66.67%

Table 2.6: Do you send electronic releases? Broken out by the number of full time employees in the technology transfer office in 2009.

# of Employees	Yes	No
3 or fewer	50.00%	50.00%
4-7	91.67%	8.33%
8-13	71.43%	28.57%
14 or more	87.50%	12.50%

Question Twenty-four: Do you send paper releases?

Table 2.7: Do you send paper releases?

	Yes	No
Entire Sample	32.43%	67.57%

Table 2.8: Do you send paper releases? Broken out by country.

Country	Yes	No
USA	28.57%	71.43%
Other	44.44%	55.56%

Table 2.9: Do you send paper releases? Broken out by enrollment.

Enrollment	Yes	No
Under 10,000	37.50%	62.50%
10,000 to 19,999	30.77%	69.23%
20,000 to 29,999	12.50%	87.50%
30,000 and above	25.00%	75.00%
Hospital	75.00%	25.00%

Table 2.10: Do you send paper releases? Broken out by technology area.

Technology Area	Yes	No
Energy	50.00%	50.00%
Medical	26.09%	73.91%
Engineering and Informatics	42.86%	57.14%
Mining, Materials & Agriculture	33.33%	66.67%

Table 2.11: Do you send paper releases? Broken out by what the technology transfer office serves.

Institution	Yes	No
A specialized unit or college such as a medical school or engineering college	50.00%	50.00%
Major research university	25.93%	74.07%
A college or university that is not a major research university	50.00%	50.00%

Table 2.12: Do you send paper releases? Broken out by the number of full time employees in the technology transfer office in 2009.

# of Employees	Yes	No
3 or fewer	40.00%	60.00%
4-7	50.00%	50.00%
8-13	28.57%	71.43%
14 or more	0.00%	100.00%

Question Twenty-five: Have you used e-marketing extensively? Google ads, blog marketing, Facebook marketing, opt in email, search engine optimization or ezine advertising? If you have used any of these means what have been the results?

1. No
2. Flintbox to market technologies.
3. Several search engines, including our own, MATTO, etc. Very light returns.
4. Not yet
5. No. We find these are ineffective means for marketing technologies.
6. Do not use.
7. Yes
8. No, but we are looking in to this.
9. Just don't have the time/resource to do all these things.
10. Use twitter, Facebook, Linkedin, etc., results are so-so.
11. Have RSS feeds for available technology; based on Google analytics, identified that many database hits result from search engine queries related to faculty - created 300 Widgets for faculty personal webpages that link to their specific technology available for licensing on our website.
12. Own website and targeted flyers.
13. Frequent use of e-mail, occasional use of social networking, primarily LinkedIn and Twitter. No tangible results to date.

Question Twenty-six: How much did the office spend on mass mailings in 2010?

Table 2.13: How much did the office spend on mass mailings in 2010? In US$

	Mean	Median	Minimum	Maximum
Entire Sample	$373.60	$0.00	$0.00	$3,240.00

Table 2.14: How much did the office spend on mass mailings in 2010? Broken out by country.

Country	Mean	Median	Minimum	Maximum
USA	$182.35	$0.00	$0.00	$2,000.00
Other	$780.00	$0.00	$0.00	$3,240.00

Table 2.15: How much did the office spend on mass mailings in 2010? Broken out by enrollment.

Enrollment	Mean	Median	Minimum	Maximum
Under 10,000	$500.00	$0.00	$0.00	$2,000.00
10,000 to 19,999	$0.00	$0.00	$0.00	$0.00
20,000 to 29,999	$1,056.67	$550.00	$0.00	$3,240.00
30,000 and above	$0.00	$0.00	$0.00	$0.00
Hospital	$0.00	$0.00	$0.00	$0.00

Table 2.16: How much did the office spend on mass mailings in 2010? Broken out by technology area.

Technology Area	Mean	Median	Minimum	Maximum
Energy	$1,333.33	$2,000.00	$0.00	$2,000.00
Medical	$289.33	$0.00	$0.00	$3,240.00
Engineering and Informatics	$0.00	$0.00	$0.00	$0.00
Mining, Materials & Agriculture	$500.00	$500.00	$0.00	$1,000.00

Table 2.17: How much did the office spend on mass mailings in 2010? Broken out by what the technology transfer office serves.

Institution	Mean	Median	Minimum	Maximum
A specialized unit or college such as a medical school or engineering college	$0.00	$0.00	$0.00	$0.00
Major research university	$386.32	$0.00	$0.00	$3,240.00
A college or university that is not a major research university	$666.67	$0.00	$0.00	$2,000.00

Table 2.18: How much did the office spend on mass mailings in 2010? Broken out by the number of full time employees in the technology transfer office in 2009.

# of Employees	Mean	Median	Minimum	Maximum
3 or fewer	$0.00	$0.00	$0.00	$0.00
4-7	$510.00	$0.00	$0.00	$2,000.00
8-13	$333.33	$0.00	$0.00	$1,000.00
14 or more	$648.00	$0.00	$0.00	$3,240.00

Question Twenty-seven: How much did the office spend on web advertising (Google ads/Facebook/other online marketing) in 2010?

Table 2.19: How much did the office spend on web advertising (Google ads/Facebook/other online marketing) in 2010? In US$

	Mean	Median	Minimum	Maximum
Entire Sample	$83.33	$0.00	$0.00	$2,000.00

Table 2.20: How much did the office spend on web advertising (Google ads/Facebook/other online marketing) in 2010? Broken out by country.

Country	Mean	Median	Minimum	Maximum
USA	$0.00	$0.00	$0.00	$0.00
Other	$250.00	$0.00	$0.00	$2,000.00

Table 2.21: How much did the office spend on web advertising (Google ads/Facebook/other online marketing) in 2010? Broken out by enrollment.

Enrollment	Mean	Median	Minimum	Maximum
Under 10,000	$333.33	$0.00	$0.00	$2,000.00
10,000 to 19,999	$0.00	$0.00	$0.00	$0.00
20,000 to 29,999	$0.00	$0.00	$0.00	$0.00
30,000 and above	$0.00	$0.00	$0.00	$0.00
Hospital	$0.00	$0.00	$0.00	$0.00

Table 2.22: How much did the office spend on web advertising (Google ads/Facebook/other online marketing) in 2010? Broken out by technology area.

Technology Area	Mean	Median	Minimum	Maximum
Energy	$666.67	$0.00	$0.00	$2,000.00
Medical	$0.00	$0.00	$0.00	$0.00
Engineering and Informatics	$0.00	$0.00	$0.00	$0.00
Mining, Materials & Agriculture	$0.00	$0.00	$0.00	$0.00

Table 2.23: How much did the office spend on web advertising (Google ads/Facebook/other online marketing) in 2010? Broken out by what the technology transfer office serves.

Institution	Mean	Median	Minimum	Maximum
A specialized unit or college such as a medical school or engineering college	$0.00	$0.00	$0.00	$0.00
Major research university	$0.00	$0.00	$0.00	$0.00
A college or university that is not a major research university	$666.67	$0.00	$0.00	$2,000.00

Table 2.24: How much did the office spend on web advertising (Google ads/Facebook/other online marketing) in 2010? Broken out by the number of full time employees in the technology transfer office in 2009.

# of Employees	Mean	Median	Minimum	Maximum
3 or fewer	$0.00	$0.00	$0.00	$0.00
4-7	$222.22	$0.00	$0.00	$2,000.00
8-13	$0.00	$0.00	$0.00	$0.00
14 or more	$0.00	$0.00	$0.00	$0.00

Question Twenty-eight: How much did the office spend on printing of promotional materials in 2010? In US$

Table 2.25: How much did the office spend on printing of promotional materials in 2010?

	Mean	Median	Minimum	Maximum
Entire Sample	$3,242.40	$1,000.00	$0.00	$16,190.00

Table 2.26: How much did the office spend on printing of promotional materials in 2010? Broken out by country.

Country	Mean	Median	Minimum	Maximum
USA	$2,274.71	$500.00	$0.00	$15,000.00
Other	$5,298.75	$3,750.00	$0.00	$16,190.00

Table 2.27: How much did the office spend on printing of promotional materials in 2010? Broken out by enrollment.

Enrollment	Mean	Median	Minimum	Maximum
Under 10,000	$5,992.86	$5,000.00	$250.00	$15,000.00
10,000 to 19,999	$1,602.22	$1,400.00	$0.00	$8,000.00
20,000 to 29,999	$4,898.00	$2,000.00	$0.00	$16,190.00
30,000 and above	$0.00	$0.00	$0.00	$0.00
Hospital	$66.67	$50.00	$0.00	$150.00

Table 2.28: How much did the office spend on printing of promotional materials in 2010? Broken out by technology area.

Technology Area	Mean	Median	Minimum	Maximum
Energy	$1,006.67	$1,000.00	$20.00	$2,000.00
Medical	$4,033.75	$1,450.00	$0.00	$16,190.00
Engineering and Informatics	$1,700.00	$0.00	$0.00	$8,000.00
Mining, Materials & Agriculture	$5,000.00	$5,000.00	$5,000.00	$5,000.00

Table 2.29: How much did the office spend on printing of promotional materials in 2010? Broken out by what the technology transfer office serves.

Institution	Mean	Median	Minimum	Maximum
A specialized unit or college such as a medical school or engineering college	$66.67	$50.00	$0.00	$150.00
Major research university	$3,907.78	$1,450.00	$0.00	$16,190.00
A college or university that is not a major research university	$2,630.00	$1,250.00	$20.00	$8,000.00

Table 2.30: How much did the office spend on printing of promotional materials in 2010? Broken out by the number of full time employees in the technology transfer office in 2009.

# of Employees	Mean	Median	Minimum	Maximum
3 or fewer	$2,674.29	$150.00	$0.00	$10,000.00
4-7	$1,255.56	$1,000.00	$0.00	$5,000.00
8-13	$3,987.50	$3,125.00	$0.00	$9,700.00
14 or more	$7,018.00	$1,500.00	$1,000.00	$16,190.00

Question Twenty-nine: How much did the office spend on trade shows/fairs/invention fairs – travel and lodging costs in 2010? In US$

Table 2.31: How much did the office spend on trade shows/fairs/invention fairs – travel and lodging costs in 2010?

	Mean	Median	Minimum	Maximum
Entire Sample	$9,054.81	$3,240.00	$0.00	$80,000.00

Table 2.32: How much did the office spend on trade shows/fairs/invention fairs – travel and lodging costs in 2010? Broken out by country.

Country	Mean	Median	Minimum	Maximum
USA	$7,868.42	$5,000.00	$0.00	$50,000.00
Other	$11,872.50	$2,870.00	$0.00	$80,000.00

Table 2.33: How much did the office spend on trade shows/fairs/invention fairs – travel and lodging costs in 2010? Broken out by enrollment.

Enrollment	Mean	Median	Minimum	Maximum
Under 10,000	$6,071.43	$5,000.00	$1,000.00	$20,000.00
10,000 to 19,999	$3,874.00	$2,870.00	$0.00	$12,000.00
20,000 to 29,999	$19,648.00	$3,240.00	$0.00	$80,000.00
30,000 and above	$25000.00	$25000.00	$0.00	$50,000.00
Hospital	$5,000.00	$3,000.00	$0.00	$12,000.00

Table 2.34: How much did the office spend on trade shows/fairs/invention fairs – travel and lodging costs in 2010? Broken out by technology area.

Technology Area	Mean	Median	Minimum	Maximum
Energy	$5,666.67	$5,000.00	$0.00	$12,000.00
Medical	$12,602.35	$4,000.00	$0.00	$80,000.00
Engineering and Informatics	$648.00	$0.00	$0.00	$3,240.00
Mining, Materials & Agriculture	$5,000.00	$5,000.00	$5,000.00	$5,000.00

Table 2.35: How much did the office spend on trade shows/fairs/invention fairs – travel and lodging costs in 2010? Broken out by what the technology transfer office serves.

Institution	Mean	Median	Minimum	Maximum
A specialized unit or college such as a medical school or engineering college	$5,000.00	$3,000.00	$0.00	$12,000.00
Major research university	$10,562.00	$3,620.00	$0.00	$80,000.00
A college or university that is not a major research university	$4,560.00	$4,120.00	$0.00	$10,000.00

Table 2.36: How much did the office spend on trade shows/fairs/invention fairs – travel and lodging costs in 2010? Broken out by the number of full time employees in the technology transfer office in 2009.

# of Employees	Mean	Median	Minimum	Maximum
3 or fewer	$1,462.86	$0.00	$0.00	$4,000.00
4-7	$8,727.27	$5,000.00	$0.00	$50,000.00
8-13	$2,3875.00	$7,250.00	$1,000.00	$80,000.00
14 or more	$8,548.00	$5,000.00	$2,500.00	$20,000.00

Question Thirty: How much did the office spend on trade shows/fairs/invention fairs – booth & attendance fees in 2010? In US$

Table 2.37: How much did the office spend on trade shows/fairs/invention fairs – booth & attendance fees in 2010?

	Mean	Median	Minimum	Maximum
Entire Sample	$6,874.23	$1,625.00	$0.00	$80,000.00

Table 2.38: How much did the office spend on trade shows/fairs/invention fairs – booth & attendance fees in 2010? Broken out by country.

Country	Mean	Median	Minimum	Maximum
USA	$2,430.56	$750.00	$0.00	$15,000.00
Other	$16,872.50	$2,870.00	$0.00	$80,000.00

Table 2.39: How much did the office spend on trade shows/fairs/invention fairs – booth & attendance fees in 2010? Broken out by enrollment.

Enrollment	Mean	Median	Minimum	Maximum
Under 10,000	$8,464.29	$2,000.00	$0.00	$44,000.00
10,000 to 19,999	$3,474.00	$2,000.00	$0.00	$15,000.00
20,000 to 29,999	$16,848.00	$1,000.00	$0.00	$80,000.00
30,000 and above	$0.00	$0.00	$0.00	$0.00
Hospital	$166.67	$0.00	$0.00	$500.00

Table 2.40: How much did the office spend on trade shows/fairs/invention fairs – booth & attendance fees in 2010? Broken out by technology area.

Technology Area	Mean	Median	Minimum	Maximum
Energy	$14,666.67	$0.00	$0.00	$44,000.00
Medical	$7,780.63	$2,000.00	$0.00	$80,000.00
Engineering and Informatics	$648.00	$0.00	$0.00	$3,240.00
Mining, Materials & Agriculture	$3,500.00	$3,500.00	$2,000.00	$5,000.00

Table 2.41: How much did the office spend on trade shows/fairs/invention fairs – booth & attendance fees in 2010? Broken out by what the technology transfer office serves.

Institution	Mean	Median	Minimum	Maximum
A specialized unit or college such as a medical school or engineering college	$166.67	$0.00	$0.00	$500.00
Major research university	$6788.95	$2000.00	$0.00	$80,000.00
A college or university that is not a major research university	$12,310.00	$2,620.00	$0.00	$44,000.00

Table 2.42: How much did the office spend on trade shows/fairs/invention fairs – booth & attendance fees in 2010? Broken out by the number of full time employees in the technology transfer office in 2009.

# of Employees	Mean	Median	Minimum	Maximum
3 or fewer	$820.00	$0.00	$0.00	$3,240.00
4-7	$6,400.00	$1,500.00	$0.00	$44,000.00
8-13	$20,812.50	$1,625.00	$0.00	$80,000.00
14 or more	$5,148.00	$3,240.00	$0.00	$15,000.00

Question Thirty-one: How much did the office spend on space ads in publications in 2010? In US$

Table 2.43: How much did the office spend on space ads in publications in 2010?

	Mean	Median	Minimum	Maximum
Entire Sample	$985.00	$0.00	$0.00	$6,000.00

Table 2.44: How much did the office spend on space ads in publications in 2010? Broken out by country.

Country	Mean	Median	Minimum	Maximum
USA	$406.25	$0.00	$0.00	$4,000.00
Other	$2,142.50	$1,500.00	$0.00	$6,000.00

Table 2.45: How much did the office spend on space ads in publications in 2010? Broken out by enrollment.

Enrollment	Mean	Median	Minimum	Maximum
Under 10,000	$1,583.33	$250.00	$0.00	$6,000.00
10,000 to 19,999	$1,026.67	$0.00	$0.00	$4,000.00
20,000 to 29,999	$980.00	$0.00	$0.00	$4,900.00
30,000 and above	$0.00	$0.00	$0.00	$0.00
Hospital	$0.00	$0.00	$0.00	$0.00

Table 2.46: How much did the office spend on space ads in publications in 2010? Broken out by technology area.

Technology Area	Mean	Median	Minimum	Maximum
Energy	$1,000.00	$0.00	$0.00	$3,000.00
Medical	$1,064.29	$0.00	$0.00	$6,000.00
Engineering and Informatics	$648.00	$0.00	$0.00	$3,240.00
Mining, Materials & Agriculture	$1,250.00	$1,250.00	$500.00	$2,000.00

Table 2.47: How much did the office spend on space ads in publications in 2010? Broken out by what the technology transfer office serves.

Institution	Mean	Median	Minimum	Maximum
A specialized unit or college such as a medical school or engineering college	$0.00	$0.00	$0.00	$0.00
Major research university	$966.67	$0.00	$0.00	$6,000.00
A college or university that is not a major research university	$2,080.00	$3,000.00	$0.00	$3,240.00

Table 2.48: How much did the office spend on space ads in publications in 2010? Broken out by the number of full time employees in the technology transfer office in 2009.

# of Employees	Mean	Median	Minimum	Maximum
3 or fewer	$462.86	$0.00	$0.00	$3,240.00
4-7	$611.11	$0.00	$0.00	$3,000.00
8-13	$2,000.00	$0.00	$0.00	$6,000.00
14 or more	$1,780.00	$0.00	$0.00	$4,900.00

Question Thirty-two: How much did the office spend on website development in 2010? In US$

Table 2.49: How much did the office spend on video and media productions in 2010?

	Mean	Median	Minimum	Maximum
Entire Sample	$2,797.20	$0.00	$0.00	$16,190.00

Table 2.50: How much did the office spend on video and media productions in 2010? Broken out by country.

Country	Mean	Median	Minimum	Maximum
USA	$1,805.56	$0.00	$0.00	$15,000.00
Other	$5,347.14	$3,240.00	$0.00	$16,190.00

Table 2.51: How much did the office spend on video and media productions in 2010? Broken out by enrollment.

Enrollment	Mean	Median	Minimum	Maximum
Under 10,000	$2,000.00	$0.00	$0.00	$8,000.00
10,000 to 19,999	$3,774.00	$1,750.00	$0.00	$15,000.00
20,000 to 29,999	$4,547.50	$1,000.00	$0.00	$16,190.00
30,000 and above	$0.00	$0.00	$0.00	$0.00
Hospital	$0.00	$0.00	$0.00	$0.00

Table 2.52: How much did the office spend on video and media productions in 2010? Broken out by technology area.

Technology Area	Mean	Median	Minimum	Maximum
Energy	$2,666.67	$0.00	$0.00	$8,000.00
Medical	$2,846.00	$1,000.00	$0.00	$16,190.00
Engineering and Informatics	$848.00	$0.00	$0.00	$3,240.00
Mining, Materials & Agriculture	$7,500.00	$7,500.00	$0.00	$15,000.00

Table 2.53: How much did the office spend on video and media productions in 2010? Broken out by what the technology transfer office serves.

Institution	Mean	Median	Minimum	Maximum
A specialized unit or college such as a medical school or engineering college	$0.00	$0.00	$0.00	$0.00
Major research university	$2,982.78	$500.00	$0.00	$16,190.00
A college or university that is not a major research university	$4,060.00	$4,120.00	$0.00	$8,000.00

Table 2.54: How much did the office spend on video and media productions in 2010? Broken out by the number of full time employees in the technology transfer office in 2009.

# of Employees	Mean	Median	Minimum	Maximum
3 or fewer	$605.71	$0.00	$0.00	$3,240.00
4-7	$3,150.00	$750.00	$0.00	$15,000.00
8-13	$333.33	$0.00	$0.00	$1,000.00
14 or more	$6,638.00	$5,000.00	$0.00	$16,190.00

Question Thirty-three: How much did the office spend on video and media productions in 2010? In US$

Table 2.55: How much did the office spend on video and media productions in 2010?

	Mean	Median	Minimum	Maximum
Entire Sample	$931.74	$0.00	$0.00	$16,190.00

Table 2.56: How much did the office spend on video and media productions in 2010? Broken out by country.

Country	Mean	Median	Minimum	Maximum
USA	$125.00	$0.00	$0.00	$1,500.00
Other	$2,775.71	$0.00	$0.00	$16,190.00

Table 2.57: How much did the office spend on video and media productions in 2010? Broken out by enrollment.

Enrollment	Mean	Median	Minimum	Maximum
Under 10,000	$250.00	$0.00	$0.00	$1,500.00
10,000 to 19,999	$467.50	$0.00	$0.00	$3,240.00
20,000 to 29,999	$3,238.00	$0.00	$0.00	$16,190.00
30,000 and above	$0.00	$0.00	$0.00	$0.00
Hospital	$0.00	$0.00	$0.00	$0.00

Table 2.58: How much did the office spend on video and media productions in 2010? Broken out by technology area.

Technology Area	Mean	Median	Minimum	Maximum
Energy	$0.00	$0.00	$0.00	$0.00
Medical	$1,399.23	$0.00	$0.00	$16,190.00
Engineering and Informatics	$648.00	$0.00	$0.00	$3,240.00
Mining, Materials & Agriculture	$0.00	$0.00	$0.00	$0.00

Table 2.59: How much did the office spend on video and media productions in 2010? Broken out by what the technology transfer office serves.

Institution	Mean	Median	Minimum	Maximum
A specialized unit or college such as a medical school or engineering college	$0.00	$0.00	$0.00	$0.00
Major research university	$1,070.00	$0.00	$0.00	$16,190.00
A college or university that is not a major research university	$1,080.00	$0.00	$0.00	$3,240.00

Table 2.60: How much did the office spend on video and media productions in 2010? Broken out by the number of full time employees in the technology transfer office in 2009.

# of Employees	Mean	Median	Minimum	Maximum
3 or fewer	$462.86	$0.00	$0.00	$3,240.00
4-7	$0.00	$0.00	$0.00	$0.00
8-13	$0.00	$0.00	$0.00	$0.00
14 or more	$3,638.00	$500.00	$0.00	$16,190.00

Chapter Three: Partnerships

Question Thirty-four: Does the office partner with any foreign universities? If so, how many in 2009?

Table 3.1: Does the office partner with any foreign universities? If so, how many in 2009?

	Mean	Median	Minimum	Maximum
Entire Sample	1.78	0.00	0.00	15.00

Table 3.2: Does the office partner with any foreign universities? If so, how many in 2009? Broken out by country.

Country	Mean	Median	Minimum	Maximum
USA	0.27	0.00	0.00	1.00
Other	4.63	3.00	0.00	15.00

Table 3.3: Does the office partner with any foreign universities? If so, how many in 2009? Broken out by enrollment.

Enrollment	Mean	Median	Minimum	Maximum
Under 10,000	2.57	1.00	0.00	15.00
10,000 to 19,999	0.29	0.00	0.00	1.00
20,000 to 29,999	5.25	5.50	0.00	10.00
30,000 and above	0.00	0.00	0.00	0.00
Hospital	0.00	0.00	0.00	0.00

Table 3.4: Does the office partner with any foreign universities? If so, how many in 2009? Broken out by technology area.

Technology Area	Mean	Median	Minimum	Maximum
Energy	0.00	0.00	0.00	0.00
Medical	2.60	0.00	0.00	15.00
Engineering and Informatics	0.40	0.00	0.00	1.00
Mining, Materials & Agriculture	0.00	0.00	0.00	0.00

Table 3.5: Does the office partner with any foreign universities? If so, how many in 2009? Broken out by what the technology transfer office serves.

Institution	Mean	Median	Minimum	Maximum
A specialized unit or college such as a medical school or engineering college	0.00	0.00	0.00	0.00
Major research university	2.28	0.50	0.00	15.00
A college or university that is not a major research university	0.00	0.00	0.00	0.00

Table 3.6: Does the office partner with any foreign universities? If so, how many in 2009? Broken out by the number of full time employees in the technology transfer office in 2009.

# of Employees	Mean	Median	Minimum	Maximum
3 or fewer	0.33	0.00	0.00	1.00
4-7	1.38	0.00	0.00	10.00
8-13	5.25	3.00	0.00	15.00
14 or more	1.40	0.00	0.00	6.00

Question Thirty-five: Does the office partner with any foreign universities? If so, how many in 2010?

Table 3.7: Does the office partner with any foreign universities? If so, how many in 2010?

	Mean	Median	Minimum	Maximum
Entire Sample	2.17	0.00	0.00	20.00

Table 3.8: Does the office partner with any foreign universities? If so, how many in 2010? Broken out by country.

Country	Mean	Median	Minimum	Maximum
USA	0.31	0.00	0.00	1.00
Other	5.88	3.50	0.00	20.00

Table 3.9: Does the office partner with any foreign universities? If so, how many in 2010? Broken out by enrollment.

Enrollment	Mean	Median	Minimum	Maximum
Under 10,000	3.29	1.00	0.00	20.00
10,000 to 19,999	0.38	0.00	0.00	1.00
20,000 to 29,999	6.50	7.00	0.00	12.00
30,000 and above	0.00	0.00	0.00	0.00
Hospital	0.00	0.00	0.00	0.00

Table 3.10: Does the office partner with any foreign universities? If so, how many in 2010? Broken out by technology area.

Technology Area	Mean	Median	Minimum	Maximum
Energy	0.00	0.00	0.00	0.00
Medical	3.13	0.50	0.00	20.00
Engineering and Informatics	0.40	0.00	0.00	1.00
Mining, Materials & Agriculture	0.00	0.00	0.00	0.00

Table 3.11: Does the office partner with any foreign universities? If so, how many in 2010? Broken out by what the technology transfer office serves.

Institution	Mean	Median	Minimum	Maximum
A specialized unit or college such as a medical school or engineering college	0.00	0.00	0.00	0.00
Major research university	2.83	0.50	0.00	20.00
A college or university that is not a major research university	0.33	0.00	0.00	1.00

Table 3.12: Does the office partner with any foreign universities? If so, how many in 2010? Broken out by the number of full time employees in the technology transfer office in 2009.

# of Employees	Mean	Median	Minimum	Maximum
3 or fewer	0.33	0.00	0.00	1.00
4-7	1.56	0.00	0.00	12.00
8-13	6.75	3.50	0.00	20.00
14 or more	1.80	0.00	0.00	8.00

Question Thirty-six: Does the office partner with any foreign universities? If so, how many does it plan to partner with in 2011?

Table 3.13: Does the office partner with any foreign universities? If so, how many does it plan to partner with in 2011?

	Mean	Median	Minimum	Maximum
Entire Sample	2.63	0.00	0.00	25.00

Table 3.14: Does the office partner with any foreign universities? If so, how many does it plan to partner with in 2011? Broken out by country.

Country	Mean	Median	Minimum	Maximum
USA	0.38	0.00	0.00	2.00
Other	7.13	4.00	0.00	25.00

Table 3.15: Does the office partner with any foreign universities? If so, how many does it plan to partner with in 2011? Broken out by enrollment.

Enrollment	Mean	Median	Minimum	Maximum
Under 10,000	4.14	1.00	0.00	25.00
10,000 to 19,999	0.38	0.00	0.00	1.00
20,000 to 29,999	7.75	9.50	0.00	12.00
30,000 and above	0.00	0.00	0.00	0.00
Hospital	0.00	0.00	0.00	0.00

Table 3.16: Does the office partner with any foreign universities? If so, how many does it plan to partner with in 2011? Broken out by technology area.

Technology Area	Mean	Median	Minimum	Maximum
Energy	0.00	0.00	0.00	0.00
Medical	3.81	0.50	0.00	25.00
Engineering and Informatics	0.40	0.00	0.00	1.00
Mining, Materials & Agriculture	0.00	0.00	0.00	0.00

Table 3.17: Does the office partner with any foreign universities? If so, how many does it plan to partner with in 2011? Broken out by what the technology transfer office serves.

Institution	Mean	Median	Minimum	Maximum
A specialized unit or college such as a medical school or engineering college	0.00	0.00	0.00	0.00
Major research university	3.44	0.50	0.00	25.00
A college or university that is not a major research university	0.33	0.00	0.00	1.00

Table 3.18: Does the office partner with any foreign universities? If so, how many does it plan to partner with in 2011? Broken out by the number of full time employees in the technology transfer office in 2009.

# of Employees	Mean	Median	Minimum	Maximum
3 or fewer	0.50	0.00	0.00	2.00
4-7	1.56	0.00	0.00	12.00
8-13	8.25	4.00	0.00	25.00
14 or more	2.60	0.00	0.00	12.00

Question Thirty-seven: Which countries does the office partner with?

1. Mexico, Canada, U.S., Europe (several)
2. NA
3. None
4. Ireland, Germany, France, Slovenia, Romania, Bulgaria, Italy, Spain, Turkey, Poland
5. Israel
6. Australia, Canada
7. UK, Slovakia, New Zealand, Spain, Portugal, Serbia, South Africa
8. Spain, France
9. Regular interaction/partnering with Italy.
10. Canada
11. UK
12. Germany
13. US, Germany, Belgium, etc.
14. No.

Question Thirty-eight: Does the Office partner with any Industry or Company groups? How Many? What industries generally? Describe your efforts.

1. 7 (life science, energy & engineering, physics, imaging, ICT, clinical medicine, VC group).
2. Yes, many. Research oriented.
3. NA
4. Mass. Biotech Council education and training.
5. Chambers of Commerce – whole range of industries access 3 chambers of commerce in the local area. Attend membership events. CEED – principally engineering, science and manufacturing companies – founding member and attend events, jamborees and development clinics on hot topics.
6. Yes. 3. Biotechnology and biomedical. We collaborate and sponsor various activities with these industry groups.
7. No
8. BIO, us and Europe.
9. 0
10. Establishing partnerships with biotech and pharma industry to further university research towards commercial development.
11. 20
12. Yes – partnerships of various forms and with various industries.
13. Primarily work with Aviation Industry.
14. 3. Mobile Communications/Carbon Nanoutibes/Satellites.
15. Ad hoc depending on circumstances, mainly technology based companies.
16. Don't know what this means.
17. BioNJ BIO LES.
18. None.
19. Yes, strategic partnerships with companies, mostly pharma.

Question Thirty-nine: Does the office support special research institutes or organizations of technology interest? If so, how? With funding? Advisory services? Explain.

1. 2 Rowett Research Institute of Health and Nutrition and Kosterlitz centre, with in kind support.
2. No
3. Yes. We are committee members/chairs for one major organization. We serve as an editor for one organization's publication.
4. Supports two State of Ohio groups: TTOC and UCAEO.
5. Yes.
6. Yes, with advisory services
7. Support to the National Institute for Aviation Research (NIAR) which is located at WSU.
8. Translation Grant Program.
9. None.
10. Yes – centres within University.

Question Forty: Does the office invest in university startup spinout companies with funding or management assistance? Does the office help arrange financing for private sector vendors in its technology?

1. Yes, with funding on occasion, and there is a relationship with an early stage VC investor.
2. Yes.
3. No.
4. Services are available but no recent spin out activity.
5. Yes to all of the above.
6. Advisory services only. We attempt to make connections to funding entities.
7. Yes; has a seed fund (CTV) and University invests in affiliated 501c3 committed to startup development.
8. Yes, we have several programs that provide anywhere from $15000 to co-investment (3:1 match required) matchmaking with potential management and investors.
9. Only in the planning stages. We may make small co-investments in prototypes based on our technologies.
10. Management assistance and help in recruiting external funding but no direct funding.
11. We have a seed fund which can put money in to spinout companies at an early stage (such as the IP costs). Due to the University's ongoing interest in many spin-outs there is also the ability to help with future funding if needed.
12. We have just spun out a start up business with an external investment of £2M and the office is investing small amounts in other spin outs.
13. Only with time, no direct $$.
14. Both startup in kind facility funding and management asst. Facilitate private sector funding when requested to by start up.
15. Yes. Designated funding available for new cos.

Question Forty-one: Continued from previous question: Is there a special investment fund with office management and employees? How many?

1. No
2. Yes, 4
3. Strategic Investment Fund open to whole University and not restricted to fund research institutes or Knowledge Exchange.
4. Yes, the University has a special fund to assist spinouts with 4-5 FTE.
5. Yes, but run by TTO (no specific employees).
6. Yes, 1
7. Yes
8. Yes – made up of existing employees.
9. There is a small budget managed by the 2 strong TT offices for the whole University.
10. Yes but managed internally
11. Yes, but through foundation

Question Forty-two: How easy is it for your office to collaborate with other university outreach departments such as the business school, grants and contracts, university outreach, international programs? Are there budget constraints to collaboration?

1. Very easy, with collaborative arrangements.
2. Very limited budget.
3. Easy to collaborate, some budget constraints.
4. Not very easy.
5. We collaborate with other business schools, including Harvard, Babson, and MIT. No budget constraints.
6. We are 'grants and contracts' and an excellent relationship exists with outreach departments as our office needs to collect, cost and approve all funding applications.
7. Fairly easy.
8. Very easy, continuous direct collaboration and support. No budgetary constraints.
9. We collaborate closely with our University Research Services, University Seed Fund
10. No budgetary constraints, but many political barriers.
11. Easy, no constraints
12. Very flexible
13. Relatively easy in areas of common interest. We do not fund one another's' efforts; so any required funding has to be allocated or secured through extra effort.
14. Very easy - we are closely affiliated with grants and contracts and the business school. Collaboration does not cost anything and is not subject to budget constraints.
15. We have an integrated research and enterprise department which has good links with these areas of the University.
16. Difficult there are only two people dealing with all enquiries and activities relating to TT
17. Fairly easy, main constraints are geographic location (we are large university) and turf wars
18. Very difficult
19. No problems identified

Question Forty-three: Has your institution merged your office with others or otherwise re-organized the institutional relationship between your office and the Office of the General Counsel, the medical or business schools, grants and contracts or other offices?

Table 3.19: Has your institution merged your office with others or otherwise re-organized the institutional relationship between your office and the Office of the General Counsel, the medical or business schools, grants and contracts or other offices?

	Yes	No
Entire Sample	24.14%	75.86%

Table 3.20: Has your institution merged your office with others or otherwise re-organized the institutional relationship between your office and the Office of the General Counsel, the medical or business schools, grants and contracts or other offices? Broken out by country.

Country	Yes	No
USA	23.81%	76.19%
Other	25.00%	75.00%

Table 3.21: Has your institution merged your office with others or otherwise re-organized the institutional relationship between your office and the Office of the General Counsel, the medical or business schools, grants and contracts or other offices? Broken out by enrollment.

Enrollment	Yes	No
Under 10,000	42.86%	57.14%
10,000 to 19,999	40.00%	60.00%
20,000 to 29,999	0.00%	100.00%
30,000 and above	0.00%	100.00%
Hospital	0.00%	100.00%

Table 3.22: Has your institution merged your office with others or otherwise re-organized the institutional relationship between your office and the Office of the General Counsel, the medical or business schools, grants and contracts or other offices? Broken out by technology area.

Technology Area	Yes	No
Energy	66.67%	33.33%
Medical	11.11%	88.89%
Engineering and Informatics	50.00%	50.00%
Mining, Materials & Agriculture	0.00%	100.00%

Table 3.23: Has your institution merged your office with others or otherwise re-organized the institutional relationship between your office and the Office of the General Counsel, the medical or business schools, grants and contracts or other offices? Broken out by what the technology transfer office serves.

Institution	Yes	No
A specialized unit or college such as a medical school or engineering college	0.00%	100.00%
Major research university	14.29%	85.71%
A college or university that is not a major research university	80.00%	20.00%

Table 3.24: Has your institution merged your office with others or otherwise re-organized the institutional relationship between your office and the Office of the General Counsel, the medical or business schools, grants and contracts or other offices? Broken out by the number of full time employees in the technology transfer office in 2009.

# of Employees	Yes	No
3 or fewer	62.50%	37.50%
4-7	10.00%	90.00%
8-13	0.00%	100.00%
14 or more	14.29%	85.71%

Made in the USA
Middletown, DE
03 November 2026

36496630R00060